Thaís Cristina Morais Vidal
Ana Beatryz P. Suzuki

Use of thinning as a way of regulating the banana harvest season

Thaís Cristina Morais Vidal
Ana Beatryz P. Suzuki

Use of thinning as a way of regulating the banana harvest season

Considerations on the Scheduled Harvest Method

ScienciaScripts

Foreword

Bananas are the most widely produced fruit in the world and are of great economic, social and nutritional importance. Most of its production is carried out by small and medium-sized family farmers who grow it in emerging and third world countries, where bananas and plantains represent not only a source of income, but also the basis of their often poor and precarious diet.

The lack of policies and incentives to encourage banana growers to become more technologically advanced so that they can make as much profit as possible from growing the fruit means that they are often held hostage by middlemen who often hold a large part of the profits generated in this chain.

This book was born out of the idea of generating information so that producers can manage their banana plantations in such a way that the harvest can be carried out at the time of their greatest interest, allowing farmers to increase their return without the need for large investments, based on the fact that the simple adoption of management techniques is capable of allowing producers to concentrate the banana harvest at times when prices are best.

We hope that the considerations in this book will be of great value to producers who want to manage their banana plantations more profitably.

CONTENTS

CHAPTER 1

INTRODUCTION

The banana *(Musa* spp.) is the most consumed fruit in Brazil and the second most produced in the country, behind only oranges (FAO, 2018; FIORAVANÇO, 2003). In 2017, Brazil produced over 6 million tonnes of bananas, making it the world's fourth largest producer of the fruit (FAO, 2018).

One of the routine practices in the management of commercial banana plantations is thinning, which consists of taking the mother plant and one or two selected shoots and systematically eliminating the rest (PEREZ et al., 1973). In Brazil, thinning is only carried out to eliminate excess shoots, thus reducing competition in the clump. However, it is this operation that ensures synchronised production in the banana plantation, and studies show that thinning can also help to regulate the time when the fruit is harvested, in order to concentrate it at a particular time of year (VIDAL, 2018; ALVES; LIMA, 2000).

In a commercial banana plantation, the production and harvesting of bunches takes place all year round, but not with the same intensity in every month, since the warm and humid climate favours the plant's development. In dry, cold weather, the banana plant reduces its activity, which means that the volume of the harvest is reduced at these times. Thus, in the Vale do Ribeira region, in the state of São Paulo, the greatest volume of production occurs between January and July and there is a shortage of the product between August and December, which results in different prices, mainly due to fluctuations in the supply of the fruit, considering that demand is constant and stable (PEREZ *et al.,* 1973).

In regions where winters are harsher, banana trees suffer various disorders caused by the lack of rain and low temperatures, such as zinc and potassium deficiencies. In places where the winter is harsher, it is common for nutritional deficiencies to be accompanied by bunch blight, caused by the near paralysis of the banana tree's metabolism, and characterised by the plant's inability to produce a floral rosette, thus causing the bunches to lose their commercial value.

Despite the reduced development and the physiological and nutritional disorders that occur in banana trees during periods of low temperatures, there is still production under these conditions, showing that there is no impediment to managing the banana plantation in such a way that the producer concentrates production at the time of their greatest interest (VIDAL, 2018).

Thus, in Ecuador and Honduras, where there is a large banana production, the orchard is managed in such a way that the bunches are harvested at a time defined by the producer, and this induction of the harvest time is carried out by thinning the shoots appropriately, and is known as the "programmed harvest" method (ALVES et al., 2014).

In Brazil, the technique of "scheduled harvesting" is practically not used by producers. However, as in other countries, thinning can be adopted not only with the aim of reducing competition in the clump, but also with the aim of determining when the bunches should be harvested, in order to allow the producer to better plan the harvest, as well as greater returns from banana growing.

CHAPTER 2

THE ECONOMIC IMPORTANCE OF BANANA GROWING

Bananas are the most produced fruit in the world and are one of the most important food sources on the planet (FAO, 2018; PERRIER et al., 2011; DANTAS et al., 1999).

According to the FAO (2018), world banana production currently stands at 113,280,302 tonnes. The Asian continent is responsible for producing 54.4% of the fruit, while the American continent is responsible for producing 25.3% of world production. In third place is the African continent, with 18.6% of production (Figure 2.1.1).

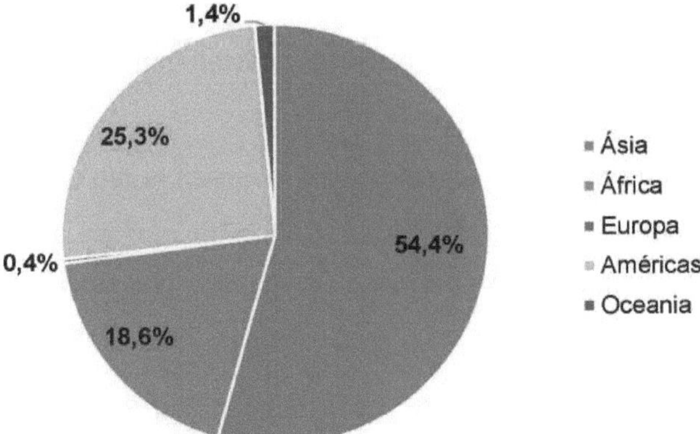

Figure 1- The continents' share of world banana production.

Source: FAO, 2018.

Brazil is the world's fourth largest banana producer, with production of 6,764,324 tonnes, behind only India (29,124,000 tonnes), China (13,324,337 tonnes) and Indonesia (7,007,125 tonnes) (FAO, 2018). However, despite the large volume of bananas produced in Brazil, the country's productivity is the lowest among the world's major banana producers. FAO data shows an average yield of 14.41 ha^{-1} in the country, while the yields of China, India and

Ecuador are 31.0, 34.4 and 36.2 t ha^{-1} , respectively. Despite this large difference, Indonesia actually has the highest average banana productivity, with a yield of 50.1 tonnes ha^{-1} (FAO, 2018).

Despite the country's low productivity, the Ministry of Agriculture points to bananas as the second most produced fruit in Brazil, and despite being the most geographically dispersed fruit in the country, the states of São Paulo and Bahia stand out as the main producers, responsible for producing 30% of the volume of bananas harvested in the national territory in the 2014/2015 harvest (MAPA, 2015). In 2017, 469,711 ha of bananas were grown and harvested in Brazil, with an average yield of 14.76 t ha^{-1} (IBGE, 2018). MAPA projections show that 7,881,000 tonnes of the fruit will be produced in 2025, which corresponds to a minimum growth of 9.1% on current production (MAPA, 2015).

However, although the volume of bananas produced in Brazil is high, the country is incipient when it comes to exports of the fruit, with almost zero expression on the international market (EMBRAPA, 2018; FIORAVANÇO, 2003).

In 2016 Ecuador was the largest exporter of the fruit, sending around 6 million tonnes of bananas to foreign markets. In the same year, Costa Rica and Guatemala were in second and third place respectively as the world's largest banana exporters, exporting around 2 million tonnes of the fruit each. In fourth place is Colombia, which exported around 1.85 million tonnes in 2016, while in the same year the Philippines was listed as the world's fifth largest exporter of bananas, at around 1.5 million tonnes (FAO, 2018).

Brazil's banana exports are far below those of the world's top five exporters, as in 2017 the country ranked fourteenth among the largest exporters of the fruit, with 41,396 tonnes of bananas exported, mainly to the European and

American markets. Thus, although Brazil has suitable climate and soil conditions for growing the fruit in a large part of its territory, these factors are not utilised, as the fruit is not of adequate quality to meet the international market (EMBRAPA, 2018; FIORAVANÇO, 2003).

In addition, there is no organisation on the part of producers or public policies that can make the Brazilian market competitive abroad, so the international market remains dominated by large multinational companies that end up controlling the stages of preparation and transport of the product in the main exporting countries, through their own production structures or in association with independent producers (FIORAVANÇO, 2003; GONÇALVES et al., 1994).

Although Brazilian exports are still insignificant on the international market, there are expectations for the expansion of this market, with projections for an increase of at least 29.9 per cent in exports by 2025, according to the Ministry of Agriculture, Livestock and Supply (MAPA, 2015).

However, before thinking about expanding the international banana market, it is necessary to know the characteristics of the production and commercialisation of the fruit in their respective regions, in order to meet regional demands and extend the fruit market beyond national borders.

In the context of the national market, in 2016 the state of São Paulo remained the largest banana producer in the country, responsible for producing 1,089,820 tonnes of the fruit on 51,512 ha, with an average yield of 21.16 t ha[-1]. The second largest producing state was Bahia, producing 1,084,548 tonnes of bananas on 72,699 ha, corresponding to a yield 70.51 % lower than that of São Paulo (EMBRAPA, 2018).

The third largest banana producer in Brazil is the state of Minas Gerais, producing 773,197 tonnes on 44,765 ha, with a yield equivalent to 173.27 t ha[-1]. In fourth place, the state of Santa Catarina is the only one of the top five producers to surpass São Paulo's production yield. In 2016, it produced 721,579 tonnes of bananas on 29,575 ha, with an average yield of 24.4 t ha^{-1} , a 15.31% higher yield than São Paulo (IBGE, 2018; EMBRAPA, 2018).

IBGE data shows that Pará is the fifth largest banana producer in the country. In 2016, the state's average yield was 11.89 tonnes of bananas per hectare, with production of 504,907 tonnes of the fruit on 42,472 ha, equivalent to 7% of national production (Figure 2) (IBGE, 2018).

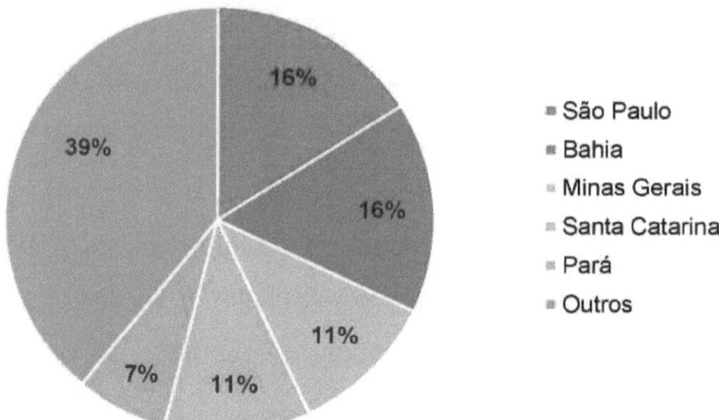

Figure 2- Participation of Brazilian states in national banana production. Source: Adapted from IBGE (2018) and EMBRAPA (2018).

CHAPTER 3

ORIGIN AND BOTANICAL CLASSIFICATION OF THE BANANA TREE

The probable centre of origin of the banana tree (Musa spp.) is the region that includes Malaysia, Indonesia, the Philippines, Borneo and Papua New Guinea (NAKASONE; PAULL, 2004). The first classification of banana trees was made by Linneo in 1735, who grouped them into the genus Musa, with the species: cavendish, sapientum, paradisiac and corniculata. However, this classification had to be replaced as there are many varieties in nature that are triploid and therefore do not fit into the classic concept of the species (SIMMONDS, 1973)

Parthenocarpic banana plants, i.e. those that produce edible fruit, originated from the wild diploid ancestors that produce seeds, Musa acuminata Colla (genome A) and Musa balbisiana Colla (genome B), through mutations and crosses. The original plants are diploids, triploids or tetraploids, which correspond to multiples of the n=11 genome, such as the AA, AAA and AAAB groups (SIMMONDS; SHEPHERD, 1955).

The various combinations of complete genomes of the parental species designated by the letters A (M. acuminata) and B (M. balbisiana) are called genomic groups (SIMMONDS; SHEPHERD, 1955). According to Dantas and Soares Filho (2000), the combination of the A and B genomes can generate three distinct chromosome levels: diploid (AA, BB and AB), triploid (AAA, AAB and ABB) and tetraploid (AAAA, AAAB, AABB and ABBB) which correspond, respectively, to two, three and four multiples of the basic number or genome of 11 chromosomes (Figure 3).

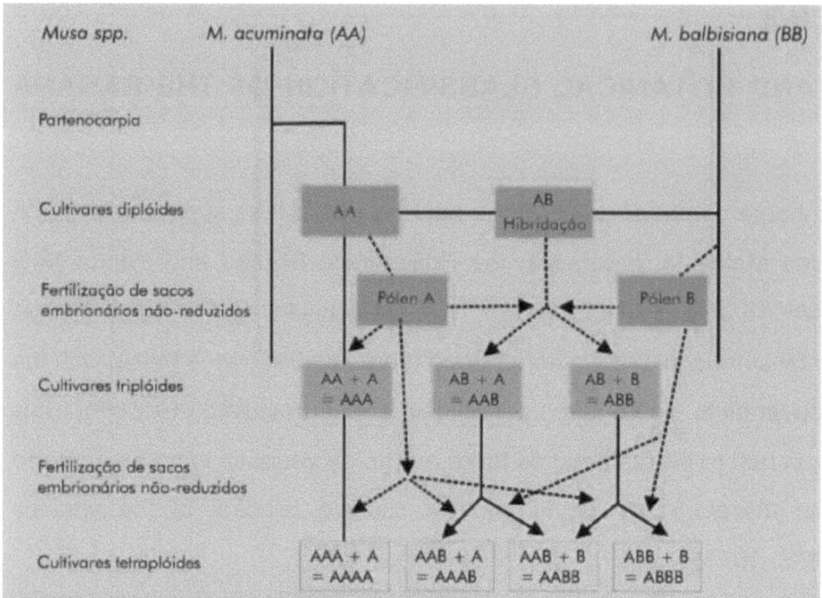

Figure 3- Evolution of banana trees.
Source: Dantas and Soares Filho (2000).

The most widespread banana varieties in Brazil are: Prata, Pacovan, Prata Anã, Maçã, Mysore, Terra and D'Angola, which belong to the AAB group and are used solely for the domestic market. The AAA genomic group, known as Cavendish, includes the Nanica, Nanicão, Grande Naine and Nanicão 'Jangada' varieties, which are accepted on the foreign and domestic markets and can be used for industrialisation. On a smaller scale, 'Ouro' (AA), 'Figo Cinza' and 'Figo Vermelho' (ABB), 'Caru Verde' and 'Caru Roxa' (AAA) are also planted. The Prata, Prata Anã and Pacovan varieties account for approximately 60 per cent of the area under banana cultivation in Brazil and are produced mainly in the Northeast, while in São Paulo and Paraná the Cavendish varieties predominate (EMBRAPA, 2010).

Main banana varieties

The genetic variability of the banana crop comes from natural mutations and somaclonal variations from *in vitro* cultivation (SONIYA et al., 2001). Sources of variability are extremely important and are used directly or indirectly in breeding programmes to develop new cultivars and select superior banana clones (HWANG; KO, 2004).

- Group AA:

The main representative of this group is the 'Ouro' cultivar.

- Group AAA:

This group includes banana plants belonging to the Cavendish subgroup, which includes the cultivars 'Nanica', 'Nanicão', 'Grande Naine' and 'Willians'; and the Gros Michel subgroup, made up of the cultivars 'Gros Michel', 'Highgate' and 'Lowgate'. This genomic group also includes the 'Caipira' cultivar.

- AAB Group:

Composed of the subgroups Prata, which includes the cultivars 'Prata', 'Branca' and 'Pacovan'; Terra, which includes the cultivars 'Terra', 'Terrinha' and D'Angola; and Figo, which includes the cultivars 'Figo Vermelho' and 'Figo Cinza'.

This genomic group also includes the cultivars 'Maçã', 'Prata-Anã' (also known as 'Enxerto') and 'Mysore'.

- AAAB Group:

Its main representatives are the cultivars 'Ouro da Mata', 'Pioneira' and 'Platina'.

CHAPTER 4

THE INFLUENCE OF CLIMATE ON THE BANANA CYCLE

The banana tree is a herbaceous plant with roots, an underground stem (called a rhizome), leaves, fruit and seeds. The primary roots originate from the rhizome, in groups of three or four, forming a fasciculated root system that can reach 5 metres horizontally. The pseudostem is formed by leaf sheaths that end in a crown of long, broad leaves with a developed centre vein. Its reproduction is vegetative and it takes 8 to 18 months for the plant to complete its cycle, the duration of which depends on genetic characteristics and soil, environmental and biotic factors, as well as the action of man (BORGES; SOUZA, 2004; MANICA et al., 1975).

According to Simão (1998), in tropical climates and irrigated areas it is possible to harvest the first bunch between 11 and 13 months after planting, while in subtropical climates and without irrigation the first harvest occurs after 15 to 18 months and around 21 to 24 months after planting the seedling in the field in colder regions.

Despite this information, it is known that the age of the banana plantation can also influence the length of the cycle and the plant's production, and it is common for the first productive cycle to be shorter and have a higher yield than the subsequent ones. However, the genotype-environment interaction must be taken into account, which produces different responses in the plants, and there may be differences not only in their size and production, but also in the number of leaves issued and the length of the cycle of banana trees belonging to the same genotype and grown in different regions of the country.

Thus, Bolfarini et al. (2014) observed that in the conditions of São Manuel, São Paulo, where the subtropical climate predominates, the 'Grande Naine', 'Prata-Anã' and 'FHIA18' banana trees took 460, 440 and 510 days from planting to harvest, respectively. In contrast, Cruz (2012) observed that, when grown in an irrigated system and in semi-arid conditions, banana trees of the same cultivars took 309, 282 and 315 days from planting to harvest, respectively, demonstrating that soil and climate conditions play a decisive role in the length of the banana cycle.

According to Manica (1997), among the edaphoclimatic factors that influence the development of banana trees, the climate plays a very decisive role in terms of the vegetative and productive cycle, the height and health of the plant, as well as the number and quality of the fruit produced. Preti (2012) points out that in the absence of ideal conditions, the plant's cycle increases due to a reduction in physiological activities, returning to normal when favourable climatic conditions are re-established.

Thus, the plasticity of the banana cycle is dependent on climatic conditions, which allow the plant to move from one phenological stage to another and thus complete its cycle. A good example is the need to accumulate 1.5 kg of leaf dry mass for flower differentiation to occur (Taulya et al., 2014), and it can be seen in the field that leaf emission is directly dependent on water availability and high temperatures, and in the absence of these conditions there is a reduction in the rate of leaf emission, which will consequently delay flowering, contributing to the expansion of the banana cycle.

The same applies to the period between the emergence of the inflorescence and the harvesting of the bunch, which, according to Manica (1997), depends on various factors, including temperature, sunshine, rain, winds, types and quantities of fertilisers, planting spacing, age of the banana plantation, number of plants per pit, cultivars, time for thinning the shoots, planting and the season.

Although the development of the bunch is influenced by various factors, the harvest time can be determined by noting the interval of days between the appearance of the inflorescence and the harvesting of the bunch, as well as observing the different indications of the approach of the harvest point, such as leaf desiccation, degree of fruit development, fragility of the flower ends, fruit diameter, fullness index, colour and consistency, as well as the resistance of the flesh to the penetrometer (Manica, 1997).

Thus, for fruit exported from Brazil, the number of days required from the appearance of the inflorescence to the harvesting of the bunch normally ranges from 70 to 90 days under ideal conditions, and in colder regions the average is 125 to 150 days. Another criterion used for determination is the degree of fruit development or angularity, by simply visualising it with a practical marker, and is used for fruit destined for the domestic market (MANICA, 1997; MANICA, 1998; BORGES; SOUZA, 2004).

Cortez (1971), working with 'Nanicão' to determine the time interval between flowering and harvest, on the coast of the state of São Paulo, observed that the shortest intervals for harvesting occur in the same period in November, December and January, on average 3 months, while the longest intervals occur in the months of May, June and July, around 5 months. When the bunches bloomed between February and April, the harvest period lengthened (125 days). This author points out that this lengthening is due to the lower amount of heat, luminosity and irradiation that are directly related to the plant's cycle. Sampaio (1978) observed in his work that some genotypes delayed flowering due to the climate of the region, which was due to the delay in leaf emission and, consequently, the number of days to flower.

Leaf emergence occurs at intervals of four days in summer and forty days in winter for temperate climate regions, while in the tropics this process takes eight to eleven days in summer and fourteen to fifteen days in winter (SUMMERVILLE, 1944).

Vegetative activity is considered very low when temperatures are below 16°C, and temperatures of around 25°C and rainfall of between 100 and 120 mm per month are what can provide excellent banana growth (CHAMPION; 1967).

Studies carried out by Sampaio (1967) in Piracicaba-SP with the Nanicão cultivar showed an average of 12.1 leaves per plant, 7.3 pencas per bunch and cycles of 289.1 days from planting to flowering. In Venezuela, Borges (1971) obtained plants that were 216 cm tall and had a 215-day cycle from planting to flowering, producing an average of 7 pencas per bunch.

Parente et al. (1981) evaluated banana varieties in the conditions of the Brazilian cerrado and found that the Prata cultivar needed 490 days to flower, with an interval of 173 days from planting to harvest. In the same study, they observed that the Maçã cultivar took 454 days to flower, and from then on it took 144 days to harvest its bunches. The 'Mysore' cultivar, on the other hand, had periods of 584 and 146 days for these two intervals, respectively.

Scarpare Filho et al. (1998) studied the first production cycle of the 'Nanicão' banana plant, analysing the performance of micropropagated seedlings, commercially available in the state of São Paulo, and conventional seedlings (pieces of rhizome, "chifrinho", "cifrão" and "guarda chuva") in terms of bunch weight, number of pieces, number of fruits and days between planting and harvest. The highest bunch weight was obtained from plants propagated via pieces of rhizome (32.1 kg), which could represent a production of 53.41 ha^{-1}, if the planting density is 1,666 plants ha^{-1}. These same seedlings had an average cycle of 483.7 days, while micropropagated seedlings took 511.9 days to harvest, and banana plants obtained from "chifrinho" type seedlings had the shortest cycle compared to the other types of seedlings (447.7 days to harvest).

Oliveira et al. (2008) reported that under the conditions of Rio Branco, AC, the 'Nanicão' cultivar showed 109, 95.8 and 90.4 days from flowering to harvest in the first, second and third cycles, respectively.

In a study carried out by Vicentini et al. (1996), the 'Grande Naine' banana tree took 330 days to flower and 130 days from flowering to harvest, totalling 460 days under the climatic conditions of Lavras-MG.

The banana cultivars 'Nanicão', 'Maçã' and 'Prata-anã', which came from micropropagated seedlings, showed a first production cycle in the municipality of Botucatu-SP of between 416 days for the 'Nanicão' cv, which was the earliest, followed by 'Prata-anã' (434 days) and 'Maçã' (437 days), as described by Leonel et al. (2004).

Scarpare Filho and Kluge (2001) observed that increasing the planting density of the 'Nanicão' banana tree increased the length of the production cycle, from 19.3 months to 19.5 months at spacings that resulted in 3,333 plants ha^{-1} and 1,333 plants ha^{-1} , respectively. In the second, third and fourth cycles, this period decreased to 10.6; 12.9; 12.8 months, when the banana plants were grown at a density of 3,333 plants ha^{-1} , and to 7.6; 9.8 and 9.0 months when the planting density was 1,333 plants ha^{-1} .

In this context, it can be said that banana trees behave in different ways, increasing or decreasing their production cycle, productivity and yield according to the local climate and the management adopted.

Among the elements of the climate that act most strongly on banana behaviour are temperature, relative humidity, wind, rain and altitude.

Temperature

The banana tree is a tropical plant that requires heat, well-distributed rainfall and high humidity for its development to be unrestricted. The optimum range for normal crop development is around 28°C (CAYÓN SALINAS, 2004), and the minimum and maximum critical temperatures are 15°C and 35°C, which are considered extreme limits for plant development (BORGES et al., 2000; MOREIRA, 1999).

In this way, studies show that below 15°C, the banana plant paralyses its development (DAMATTO JUNIOR et al., 2005; BORGES; SOUZA, 2004; BORGES et al., 2000), while above 35°C there is a loss in the plant's development rates, due to the dehydration of tissues, especially the leaves (BORGES et al., 2000). In addition, high temperatures induce the closure of stomata and reduce the photosynthetic rate, leading to a reduction or inhibition of leaf growth, similar to what occurs due to water stress (COELHO, 2009).

On the other hand, short periods of exposure to temperatures below 12°C cause damage to the banana plant's root system, hinder plant nutrition, reduce or paralyse its activity, burn the leaf surface, reduce the photosynthetic rate, delay and hinder the emission of the inflorescence and cause yellowing of the leaves (CRISOSTOMO; NAUMOV, 2009). However, despite the fact that the leaves burn due to the low temperatures, when the rhizome buds are not affected, the plant returns to activity the following spring, when the ideal conditions for its development are established. However, the absence of leaves causes the bunches to burn and darken, as well as slower development, a delay in the cycle and the production of small, deformed fruit of low commercial value (MANICA, 1997).

When temperatures below 12°C occur during the fruit filling phase, chilling occurs, characterised as a physiological disorder that damages the skin tissues, causing them to darken. Chilling can occur in subtropical regions, where the minimum night-time temperature reaches 4.5°C to 10°C. This disorder is most common in the field, but can also occur when the bunch is being transported, in the air conditioning chamber or just after the banana has turned yellow. Fruits affected by chilling have their ripening impaired and their peel blackened, with a consequent loss of commercial value (PAIVA et al., 2011).

Borges et al (2000) point out that low temperatures can cause the so-called "embuchamento" or "choking" of the bunch. This phenomenon makes the bunch unviable for commercialisation and is described by Vidal (2018) as the compaction of the floral rosette on the inside of the pseudostem, which hinders or prevents the inflorescence from being released, deforming the bunch, which often ends up causing cracks and ruptures in the pseudostem structure, due to the pressure it exerts on the inside of the stem as a result of the twisting of the stalk, which continues to develop even without the inflorescence being released (Figure 4 (A) and (B)).

Figure 4- Damage caused by the inflorescence "embuchamento". (A): Twisting of the stalk. (B): Deformation of the unissued bunch.
Source: Author, 2016.

Even in cases where this is possible, the bunch has often already started to develop even before it emerges, due to the length of time it takes for this to happen. Even in these cases, the bunch is lost because it is already deformed and fruiting is impaired (Figure 5).

Figure 5 - Deformed bunch, due to its formation inside the pseudostem.

Source: Author, 2017.

With this information, it can be seen that the banana plant is typical of tropical regions and for this reason its production is limited to the region of the globe between 30° North and South latitudes. High and uniform temperatures are essential for obtaining high yields from banana trees.

Precipitation

The banana tree is a plant with high and continuous water consumption, due to the morphology and hydration of its tissues. The highest banana yields are associated with a total annual rainfall of 1,900 mm, well distributed throughout the year, which results in an approximate demand of 160 mm per month^{-1} or 5 mm per day^{-1} (BORGES; SOUZA, 2004).

In tropical regions, during the summer period, which corresponds to the season with the highest concentration of rainfall, the banana foliage has a characteristic dark green colour and the plant develops better, forming fruit of excellent quality. However, during the winter period, when there is less rainfall, associated with lower average temperatures, banana trees develop more slowly and the foliage takes on a more yellowish colour (MANICA, 1997).

Water deficit becomes more serious in the flower differentiation and early fruiting stages, because if the plant is subjected to severe water deficiency, the leaf rosette compresses, which makes it difficult or even prevents the inflorescence from launching. As a result, the bunch can lose its commercial value (LIMA et al., 2012).

According to Borges and Souza (2004), in addition to evapotranspiration conditions, water supply is related to the type of soil. When grown on deeper soils with good moisture retention capacity, banana trees have a water demand of around 100 mm per month^{-1}. On the other hand, when grown in soils with a lower water retention capacity, 180 mm per month is required^{-1}. Regardless of the type of soil in which the crop is grown, it is of fundamental importance that the water supply ensures availability of no less than 75 per cent of the soil's water retention capacity, without the risk of saturation, which impairs aeration (BORGES; SOUZA, 2004; LIMA et al., 2012).

Manica (1997) points out that there are banana-growing regions where rainfall is scarce, but there are other places where it can rain a lot, but with poor distribution, as in some regions of Brazil; in these conditions it is necessary to irrigate in the months of low rainfall to ensure good production.

Wind

Wind is an important climatic factor, as it can cause anything from minor damage to the complete destruction of the banana plantation. Winds of less than 30 kilometres per hour^{-1} do not normally harm the plant and are not limiting for banana cultivation (MANICA, 1997). However, its intensity can cause major damage: chilling, in the case of cold winds; dehydration of the plant, as a result of high evaporation; splitting of the secondary veins of the leaves; reduction in leaf area, due to the splitting of the split leaf; breaking of roots; breaking of the plant and toppling (LIMA et al., 2012).

Losses caused by wind can reach up to 30 per cent of production, and several authors recommend choosing low-growing cultivars, such as Nanica, which are more resistant to wind than the medium-sized Nanicão and Grande Naine (MANICA, 1997; MANICA 1998, BORGES; SOUZA, 2004).

Altitude

Banana trees are grown in places where altitudes vary from 0 to 1000 metres above sea level, and altitude influences the banana tree's cycle. An example of this is the Cavendish group of banana trees which, when grown at low altitudes (0 to 300 m), have a cycle of between 8 and 10 months, and when grown in regions where the altitude is 900 metres, it takes around 18 months for the plant to complete its cycle (BORGES; SOUZA, 2004). The same authors also emphasise that for every 100 m increase in altitude, the production cycle increases by 30 to 45 days, considering the same growing conditions.

Relative humidity

Typical of tropical regions, the banana tree grows best in places with an average annual relative humidity of around 80 per cent. This condition contributes to faster and longer leaf emission, which favours inflorescence emission and uniform fruit colour (BORGES; SOUZA, 2004; LIMA et al., 2012).

However, when combined with rainfall and high temperatures, the optimum level of relative humidity contributes to an increase in the incidence of fungal diseases, particularly black and yellow Sigatokas, caused by the fungi *Mycosphaerella fijiensis* and *M. musicola*. On the other hand, low relative humidity favours leathery leaves and a shorter lifespan (BORGES; SOUZA, 2004; LIMA et al., 2012).

CHAPTER 5

THINNING BANANA TREES AND CONTROLLING THE PRODUCTION SEASON

Cultural practices, when carried out correctly, at the appropriate time and adapted to the peculiarities of the ecosystem, are essential for the development and production of the various crops. In banana management, the main recommended cultural practices are: weeding, thinning, defoliation, elimination of the male rachis ("heart"), bagging the bunch, propping, weed control and cutting the pseudostem after harvest (ALVES; LIMA, 2000).

Thinning is another routine operation in the banana plantation and is characterised by its great importance in terms of regulating production, cluster and fruit size, plant alignment and even the useful life of the plantation (LIMA et al., 2003).

It consists of eliminating excess shoots from the clump, since the banana tree produces a variable number of offspring during its productive cycle (ALVES; LIMA, 2000) and if all the shoots emitted develop, the clump will have a large number of plants that will compete with each other, not only for nutrients, but also for light and water, which will mean that the production of each individual plant will be lower and of poorer quality, especially in terms of fruit size.

Alves and Oliveira (1999) consider that the main objective of thinning is to eliminate excess shoots, with the consequent economic development and obtaining sustainable crops, by maintaining a population in the banana plantation that allows good productivity to be combined with high fruit quality, as well as promoting pest control through greater aeration of the plantation.

The thinning operation is carried out when the shoots of the clump are between 0.20 and 0.30 metres high, and these are cut with a knife, machete or sledge. After cutting, the apical buds of the respective tillers are removed with the aid of a bud puller (lurdinha) or a sledge. Removing the buds from the tillers is up to 75 per cent more efficient and effective than other methods used in thinning (LIMA et al. 2018), as the apical buds are removed in order to inhibit further sprouting of the already removed offspring.

Gonzaga Neto (1995) recommended that during orchard formation, thinning should be carried out at four, six and ten months of age. However, in other cycles, thinning is carried out systematically in commercial banana plantations and is commonly associated with defoliation (LIMA et al., 2012). Despite this, the intensity of this operation is basically conditioned by economic factors, which are mainly related to the seasonal variation in banana prices (LIMA et al. 2018). Thus, normally in times of lower prices, Brazilian banana growers choose to carry out thinning less frequently in order to reduce production costs.

During the thinning operation, it is recommended to leave the mother, one son and one grandson in each cycle of the banana plantation, eliminating the remaining shoots (Figure 7). Despite this, there are producers who choose to leave only the mother plant, or the mother and one or two followers (children), eliminating the other shoots (LIMA et al. 2018).

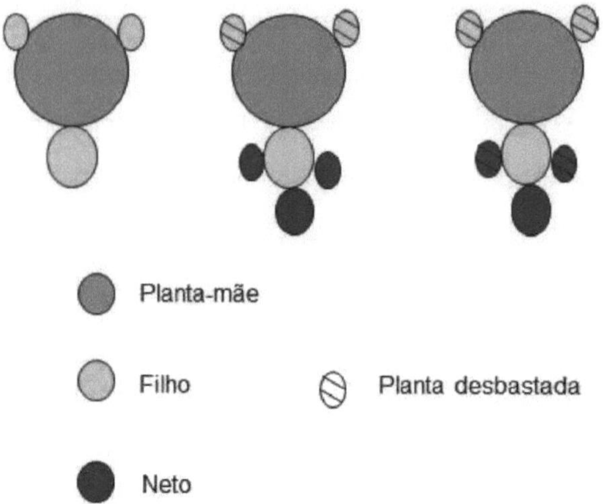

Figure 7- Scheme for thinning the banana clump, leaving the mother plant, a son and a grandson.

Source: Adapted from Neto; Melo (2018).

Regardless of the system adopted, it is recommended that growers select and keep the most vigorous and developed shoots in the clump. However, there are authors who say that the location of the shoots to be kept in the clump is not considered a factor of great importance, while other authors say that the position of the shoots kept in the clump helps to maintain the alignment of the banana plantation, at least in the first years of cultivation, thus making it easier to carry out other cultural treatments, such as weeding and fertilisation, at least during the period in which it is possible to maintain this alignment (LIMA et al., 2018; GONZAGA NETO, 1995).

Despite the importance of thinning in reducing clump competition, in several countries it has been adopted not only to eliminate clump competition, but also to regulate the production season.

Thus, according to Aubert (1971), banana production can be adjusted not only according to climatic conditions, but also according to economic conditions, based on the selection of planting material, the thinning of shoots and the planting season.

According to Rodrigues and Souza (1947), two factors play a role in the distribution of production: the selection of shoots and the duration of vegetative activity. Manica (1973), in his work, found the highest production per hectare in treatments in which all the shoots were kept in the clump, with the second cycle showing the highest production.

Harvest control in banana plants has already been successfully used in the Canary Islands, South Africa, Israel, Colombia, the Philippines, Ecuador and Central America (VALDENEBRO, 1984). In these countries, the determining factors for achieving harvest at pre-determined times were based on the time of selection and the size of the selected tillers.

In Brazil, studies aimed at determining the ideal time for thinning banana trees in the different producing regions are lacking and depend on research that evaluates not only the size and timing of tiller selection, but also the influence that local environmental conditions have on the dilation of the banana tree's phenological stages.

Thus, for the conditions of the municipality of Piracicaba-SP, Manica et al. (1973) working with the relationship between the selection of shoots at different times and the number of days from planting to flowering, obtained the result that shoots selected in May had a longer cycle with an average of 293.4 days and January shoots, conducted alone and in clumps conducted with shoots selected in March had a shorter cycle of 278.6; 273.0 and 263.8 days, respectively.

In the conditions of Cristais Paulista-SP, Flori et al. (2008) evaluated the effect of the management of the mother plant on the production and development period of the daughter plants of 'Prata-Anã' and found in their work that the shoot selection seasons of February, March and April resulted in the highest bunch weights: 23.2, 22.6 and 22 kg respectively. The lowest bunch weights were obtained in August, September and October: 15.2, 15.6 and 14.8 kg, respectively. In the same study, the authors observed that planting the banana plantation without the mother plant reduced the cycle of the daughter plant (19.9 months). In the case of banana plantations without thinning, the time taken to harvest the shoots was longer, at 21.2 months.

In 'Mysore' banana trees grown in Porto Lucena-RS, Manica et al. (1995) tested different times of shoot selection: at 4, 5, 6, 7 and 8 months after planting. The results showed that shoot selection after 7 months reduced harvest time by up to 45 days, which resulted in a total cycle of 620.97 days.

According to Alves et al. (2014), in banana plantations with a higher technological level, and those whose fruit is destined for export, the appropriate thinning should be carried out based on the phenological knowledge of the plant in the respective producing region. This knowledge makes it possible to obtain sequential production capable of meeting market demand, and the quality of the product is directly related to the number of offspring that develop on each clump.

Thinning of shoots has been carried out in almost all major banana producing countries and this practice avoids competition from the follower and helps to obtain an optimum population that allows high yields to be achieved. In addition, thinning allows producers to cater for seasonal variations in product supply (ALVES et al., 2014).

In addition to the correct adoption of the thinning technique, combined with knowledge of the behaviour of banana trees in the respective producing regions, the orchard must be set up with seedlings of a uniform size, which will allow the harvesting season to be planned in the first cycles of the banana plantation.

However, despite the use of seedlings of uniform size, it is possible to observe that among the plants that make up the orchard, some of them show differences in height and pseudostem circumference, due to the physiological age of the seedlings used. There is therefore a great diversity of criteria as to the ideal time to carry out thinning, and determining this moment depends on the climatic conditions to which the plant is subject, as well as the market situation and even timing issues (ALVES et al., 2014).

In Ecuador, a total periodic thinning system has been developed that allows the harvest time to be regulated according to the target market. Known as the "programmed harvest" method, this system is based on the principle of selecting shoots of similar ages and eliminating plants of different ages, such as those in bloom and even those with bunches, allowing bunches of the selected shoots to be harvested at the time of the best prices for the product on the market (ALVES et al., 2014).

The "scheduled harvest" method allows harvesting to take place within a maximum of 12 weeks after flowering and is carried out every nine months, depending on the demands of the product on the market (ALVES; OLIVEIRA, 1999).

Although the "programmed harvesting" method is not widespread in Brazil, the adoption of the thinning technique to regulate the harvesting season can be successfully employed in the country, enabling
not only to determine a specific harvest time, but also so that producers can increase their income by harvesting fruit in the off-season, when higher prices are paid for the fruit.

CHAPTER 6

FINAL CONSIDERATIONS

The information gathered in this book shows that the "programmed harvest" method can be successfully adopted by Brazilian banana growers, and that its main benefits are not only the prediction of the harvest time, but also the possibility of concentrating this operation in periods of the year pre-established by the producer.

Determining the concentration of the harvest time not only makes it possible to better plan banana plantation operations, but also allows farmers to increase their income by concentrating the harvest in the off-season.

However, growers should bear in mind that not only does the genotype influence the banana cycle, but that the environment also makes the phenotype responsible for extending not only the banana cycle as a whole, but also the length of its phenological phases.

Therefore, in order for the "scheduled harvest" method to be used successfully, it is essential to know the banana cultivar to be used, the local environmental conditions and the plant's behaviour under these conditions.

Biblical references

ALVARENGA, A.A.; SOARES, A.M.; ALVES, J.D.; OLIVEIRA, L.E.M.; PAIVA, R.; DELÚ FILHO, N.; CAIO, S.F. Plant ecophysiology. Lavras: UFLA/FAEPE, 2001, 69p.

ALVES, E.J. Main banana cultivars in Brazil. Revista Brasileira de Fruticultura, Jaboticabal, v. 12, p. 45-61, 1990.

ALVES, E.J.; OLIVEIRA, M. de A.; LIMA, J.L.L.; OLIVEIRA, S.L. de. Climate requirements. In: ALVES, E.J. A cultura da banana: Aspectos técnicos, socioeconômicos e agroindustriais. 2 ed. Brasília: EMBRAPA. 1999. 585 p.

ALVES, E.J.; LIMA, M.B. Chapter 11. In: CORDEIRO, Z.J.M. Banana: production and technical aspects. Brasília: Embrapa Comunicação para Transferência de Tecnologia, 2000. p.17-23. (Frutas do Brasil, 1).

ALVES, E.J.; LIMA; M.B.; CARVALHO, J.E.B. de; BORGES, A.L. Chapter VII: Cultivation and harvesting. Available at: <http://www.agencia.cnptia.embrapa.br/recursos/Livro_Banana_Cap_7ID - 3pTuregodF.pdf> Accessed on 04 August 2014.

AUBERT, Bernard. Action du climat sur le comportement du bananier en zones tropicale et subtropicale. Fruits, Paris, v. 26, p. 175-188, 1971.

AWAD, M.; CASTRO, P.R.C. Introduction to plant physiology. São Paulo: Nobel, 1983.

BELÁLCAZAR CARVAJAL, S.L. Plantain cultivation (Musa AAB Simmonds) in the tropics. Cali: ICA, 1991. 376 p.

BOLFARINI, A.C.B.; JAVARA, F.S.; LEONEL, S.; LEONEL, M. Growth, phenological cycle and production of five banana cultivars under subtropical conditions. Revista Raízes e Amidos Tropicais, Botucatu, v. 10, n. 1, p. 74-89, 2014.

BORGES, A.L. Climatic requirements. In: Banana production system for the state of Pará. Production system, 9. EMBRAPA Cassava and Fruit Growing, 2014. Available at: <https://www. infoteca.cn ptia.embrapa.br/bitstream/doc/1002115/1/Sistem aProducaoBanana.pdf> Accessed on: 01 April 2018.

BORGES, A.L.; SOUZA, L. S. **The cultivation of banana trees.** Embrapa Cassava and Fruit Growing, 2004. 279 p.

BORGES, A.L.; SOUZA, L. da S.; ALVES, E.J. Edaphoclimatic requirements. In: CORDEIRO, Z.J.M. **Banana: production and technical aspects.** Brasília: Embrapa Comunicação para Transferência de Tecnologia, 2000. p. 17-23. (Frutas do Brasil, 1).

BORGES, O.L. Comparative study of different plantain and cambur clones. **Agronomia Tropical,** v.14 p. 265-275. 1971.

CAMOLESI, M.R.; NEVES, C.S.V.J.; MARTINS, A.N.; SUGUINO, E. Phenology and productivity of banana cultivars in Assis, São Paulo. Revista Brasileira de Ciências Agrárias, Pernambuco, v.7, n. 4, p. 580-585, 2012a.

CAMOLESI, M.R.; NEVES, C.S.V.J.; MARTINS, A.N.; SUGUINO, E. Performance of banana cultivars in the region of Médio Paranapanema, São

Paulo. **Semina: Ciências Agrárias,** Londrina, v. 33, s. 1,p. 2931-2938, 2012b.

CARVALHO, A. de; MARCOLINI, C.D.M.; SOARES JÚNIOR, D.; LIRA, M.P. de; GOMES, M.S.; LLANILLO, R.F.; CARNEIRO, S.L. **Family production system in the north of Paraná: Grains and bananas.** Instituto Agronómico do Paraná (IAPAR), Instituto Paranaense de Assistência Técnica e Extensão Rural (EMATER), Londrina, 2008, 4 p. Available at: <http://www.iapar.br/arquivos/File/zip_pdf/redereferencia/SR7.pdf> Accessed on: 02 September 2016.

CAVIGLIONE, J.H.; KIHL, L.R.B.; CARAMORI, P.H.; OLIVEIRA, D. **Climatic maps of the state of Paraná.** Londrina: IAPAR, 2000. 1 CD.

CAYÓN SALINAS, D.G. Ecophysiology and productivity of the plantain *(Musa* AAB Simmonds). In: Reunión Internacional para coperación en la investigación de banano en el Caribe y en América Central, 16. 2004, Oaxaca, México. **Memórias...** San José, Costa Rica: CORBANA, 2004. p.172- 183.

CEASA-PR, SUPPLY CENTRES OF PARANÁ. **Daily price quotation.** Available at: <http://www.ceasa.pr.gov.br/modules/conteudo/conteudo.php?conteudo= 386> Accessed on: 05 January 2018

CHAMPION, J. 1967. **Bananas and their Culture. Institut Français de Recherches Fruitières** Outre-Mer (IFAC)/Editions SETCO. Paris

COELHO, E.F. **Course on Irrigated Banana Growing.** Document 176. Cruz das Almas: EMBRAPA, 2009.

COELHO JÚNIOR, L.M. Regional concentration of the gross value of banana production in Paraná, Brazil (1995 to 2010). **Ciência Rural,** Santa Maria, v. 43, n. 12, p. 2304-2310, 2013.

CORTEZ, J.V. Observations from flowering to harvest: Banana cultivation - *Musa* Cavendish cultivar Nanicão. In: Congresso brasileiro de fruticultura, 1, 1970, Campinas, **Anais...** Campinas: Sociedade Brasileira de Fruticultura, 1971.

CRISÓSTOMO, L. A.; NAUMOV, A. **Fertilising for High Productivity and Quality of Tropical Fruit Trees in Brazil.** International Potash Institute Bulletin, 18. Fortaleza: Embrapa Agroindústria Tropical, 2009. 238 p.

DAMATTO JÚNIOR, E.R.; CAMPOS, A.J. de; MANOEL, L.; MOREIRA, G.C.;

LEONEL, S,; EVANGELISTA, R.M. Production and characterisation of 'Prata-anã' and 'Prata-zulu' banana fruits. **Revista Brasileira de Fruticultura,** Jaboticabal, v. 27, n. 3, p. 440-443, 2005.

DANTAS, J.L.L.; SHEPHERD, K.; OLIVEIRA e SILVA, S. de; FILHO, W. dos S.S. Botanical classification, origin, evolution and geographical distribution. In: ALVES, E.J. **A cultura da banana: Aspectos técnicos, socioeconômicos e agroindustriais.** 2 ed. Brasília: EMBRAPA. 1999. 585 p.

DANTAS, J.L.L.; SOARES FILHO, W. dos S.S. Botanical classification, origin and evolution. In: **Banana Production 1**: Technical aspects. 1 ed. Brasília: Embrapa Comunicação para Transferência de tecnologia, 2000. p. 12-16.

DEBIASI, C. **Physiological and biochemical characterisation of apical dominance in banana *(Musa acuminata* Colla)**. Thesis (Doctorate) - Universidade Estadual Paulista, Faculdade de Ciências Agronômicas, Botucatu, 2007. 154 f.

DONATO, S.L.R.; SILVA, S. de O. e; PASSOS, A.R.; NETO, F.P.L.; LIMA, M.B. Evaluation of banana hybrid varieties under irrigation. **Revista brasileira de Fruticultura**, v. 25, n. 2, p. 348-351, 2003.

DONATO, S.L.R.; ARANTES, A. de M.; COELHO, E.F.; RODRIGUES, M.G.V. Ecophysiological considerations and management strategies for banana trees. In: VIII Brazilian Symposium on Banana Growing. Montes Claros. Lectures and abstracts...Belo Horizonte: Epamig, 2015. 1 CD-ROM.

EMBRAPA, BRAZILIAN AGRICULTURAL RESEARCH COMPANY. Brazilian banana production in 2016. Available at: <http://www.cnpmf.embrapa.br/Base_de_Dados/index_pdf/dados/brasil/b anana/b1_banana.pdf> Accessed on: 21 July 2018.

EMBRAPA, BRAZILIAN AGRICULTURAL RESEARCH COMPANY. Destinations of Brazilian banana exports in 2017. Available at: <http://www.cnpmf.embrapa.br/Base_de_Dados/index_pdf/dados/brasil/b anana/b71_banana.pdf>

EMBRAPA, BRAZILIAN AGRICULTURAL RESEARCH COMPANY. Organic production system for banana cultivation. September 2010. Available at: <https://sistemasdeproducao.cnptia.embrapa.br/FontesHTML/Banana/Sis temaOrganicoCultivoBanana/anexos.htm> Accessed on 16 August 2016.

EMBRAPA, BRAZILIAN AGRICULTURAL RESEARCH COMPANY. Irrigated Banana Production System. Production Systems 4, 2009. Available at: <https://sistemasdeproducao.cnptia.embrapa.br/FontesHTML/Banana/Ba naneiraIrrigada/cultivares.htm>. Accessed on 16 August 2016.

BRAZILIAN AGRICULTURAL RESEARCH COMPANY - EMBRAPA. Brazilian soil classification system. 3.ed. Brasília, 2013. 353p.

FAO. Food and Agriculture Organisation. Available at: <http://faostat.fao.org/site/339/default.aspx>. Accessed on: 03 August 2014.

FOOD AND AGRICULTURE ORGANISATION OF THE UNITED NATIONS - FAO. Available at: <http://www.fao.org/faostat/en/#home> Accessed on 05 January 2018.

FERNANDEZ-CALDAS, E. et al. Foliar analysis of the plantain in two phases of its development: flowering and cutting. Fruits, Paris, v. 32, n. 11, p. 665-671, 1977.

FERREIRA DE CASTRO, C.E.; MAY, A.; GONÇALVES, C. Heliconia species as cut flowers. **Revista Brasileira de Horticultura Ornamental**, Campinas, v. 12, n. 2, p. 87-96, 2007.

FIORAVANÇO J.C. World banana market: production, trade and Brazilian participation. **Economic Information**, v. 33, n. 10, p. 15-27, 2003.

FLORI, J.E.; SCARPARE FILHO, J.A.; RESENDE, G.M. de. Evaluation of the production cycle of the daughter plant as a function of the management of the mother plant **at different times of the year in the 'Prata-anâ' banana plant. Ciência e Agrotecnologia**, Lavras, v. 32, n. 3, p. 969-973. 2008.

FREITAS, W. da S.; RAMOS, M.M.; COSTA, S.L. da. Irrigation demand for banana crops in the São Francisco River basin. **Revista Brasileira de Engenharia Agrícola e Ambiental**, Campo Grande, v. 2, n. 4, p. 343-349, 2008.

GONÇALVES, J.S.; PEREZ, L.H.; SOUZA, S.A.M. International and banana production: the productive and commercial structure of the world banana complex. **Agricultura em São Paulo**, v. 41, n. 3, p. 161-188, 1994.

GONZAGA NETO, L. **Banana cultivation**. 1995. Available at: <https://core.ac.uk/download/pdf/33890291.pdf> Accessed on: 16 July 2018.

GREEN, G.C.; KUHNE, F.A. Growth of the banana plant in relation to winter air temperature fluctuations. **Agroplantae**, Pretoria, v. 1, p. 157162, 1960.

GREEN, G.C.; KUHNE, F.A. Research note. The response of banana foliar growth to widely fluctuating air temperatures. **Agroplantae**, Pretoria. v.2, p. 105-107, 1970.

HASSELO, R.N. Evaluation of the circumference of the pseudostem as in grown index for the Gross Michel banana. **Tropical Agriculture**, Trindad, v. 39, n. 1, p. 57- 63, 1962.

HOLDER, G.D.; CUMBS, F.A. Effects of water supply during floral initiation and differentiation on female flower production by robusta banana. **Experimental Agriculture**, New York, v. 18, n. 2, p. 183-93, 1982.

HWANG, S.C.; KO, W.H. Cavendish banana cultivars resistant to *Fusarium* wilt acquired through somaclonal variation in Taiwan. **Plant Disease**, Davis, v.88, p.580-588, 2004.

IBGE - Brazilian Institute of Geography and Statistics. Agricultural production indicators. 2018. Available at: <ftp://ftp.ibge.gov.br/Producao_Agricola/Fasciculo_Indicadores_IBGE/est ProdAgr_201701.pdf>. Accessed on: 05 February 2018.

AGRONOMIC INSTITUTE OF PARANÁ - IAPAR. Climate maps of Paraná. Available at: <http://www.iapar.br/modules/conteudo/conteudo.php?conteudo=597> Accessed on: 02 January 2018.

IAPAR, AGRONOMIC INSTITUTE OF PARANÁ. Zoning Paraná state agriculture. Available at: <http://www.iapar.br/modules/conteudo/conteudo.php?conteudo=1095>. Accessed on: 17 August 2016.

INMET. NATIONAL METEOROLOGICAL INSTITUTE. Precipitation and relative humidity database. Available at: < http://www.inmet.gov.br/sim/gera_graficos.php>. Accessed on: 17 August 2018.

ITCG. LAND, CARTOGRAPHY AND GEOSCIENCES INSTITUTE. Climate - State of Paraná. Available at: <http://www.itcg.pr.gov.br/arquivos/File/Produtos_DGEO/Mapas_ITCG/P DF/Mapa_Climas_A3.pdf> Accessed on: 17 August 2018.

LASSOUDIERE, A. Quelques aspects de la croissance et du développement du bananier "Poyo" en Cote d'Ivoire. III. Le faux-tronc et le système foliaire. Fruits, Paris, v. 33, n. 6, p. 373-412, 1978.

LEONEL, S.; GOMES, E.M.; PEDROSO, C.J.. Agronomic performance of micropropagated banana plants in Botucatu-SP. Revista Brasileira de Fruticultura, Jaboticabal, v. 26, n. 2, p. 245-248. 2004.

LIMA, M.B.; ALVES, E.L.; SILVEIRA, J.R.S. Cultural practices. In: Banana: the producer asks, Embrapa answers. Collection 500 questions, 500 answers. Embrapa cassava and fruit growing. Brasília, DF: Embrapa Information Technology, 2003. 182 p.

LIMA, M.B.; SILVA, S.O.; JESUS, O.N.; OLIVEIRA, W.S.J.; GARRIDO, M.S.; AZEVEDO, R.L. Evaluation of banana cultivars and hybrids in the Recôncavo Baiano. Ciência e Agrotecnologia, Lavras, v.29, n.3, p.515- 520, 2005.

LIMA, M.B.; SILVA, S.O.; FERREIRA, C.F. **Banana:** the producer asks, Embrapa answers. 2. ed. Brasília, DF: Embrapa, 2012. 214 p.

LIMA, J.R.; MORAES, W. da S. SILVA, S.H.M.G. da. Physiological responses in banana seedlings treated with strobirulins. **Semina: Ciências Agrárias**, Londrina, v. 33, n. 1, p. 77-86, 2012.

LIMA, M.B.; ALVES, E.J.; CARVALHO, J.E.B. de; BORGES, A.L. Banana: Thinning. EMBRAPA Information Agency. Available at: <http://www.agencia.cnptia.embrapa.br/Agencia40/AG01/arvore/AG01_5 _41020068054.html> Accessed on: 16 July 2018.

MANICA, I.; SIMÃO, S.; OTTO, C. K.; PEREZ, F. P. Z. CONDE, R. A. Influence of the time of shoot selection on the development and production of the first shoot (second cycle) of the banana tree *(Musa acuminata)* **cv. 'Nanicão'. Revista Ceres**, Viçosa, v.22. p. 359 -366. 1973.

MANICA, I.; CARVALHO, R. I. N.; KIST, H.; VIONE, G. F.; BARRRADAS, C. I. N. Effect of shoot selection times on the cycle and production of mother banana plants *(Musa acuminata)* cultivar Mysore in Porto Lucena, RS. **Pesquisa Agropecuária** Brasileira, Brasília, v. 30, n. 6 p. 825-829. 1995.

MANICA, I. **Fruticultura Tropical 4:** banana. Porto Alegre: Cinco continentes, 1997. 485 p.

MANICA, I. **Bananas:** from planting to ripening. Porto Alegre: Cinco continentes, 1998. 99 p.

MATTHIESEN, M.L.; BOTEON, M. **Analysis of the main banana producing centres in Brazil**. 2003. Available at: <http://www.cepea.esalq.usp.br/pdf/banana.pdf> Accessed on: 02 September

2018.

MAP, MINISTRY OF AGRICULTURE, LIVESTOCK AND SUPPLY. **Agribusiness Projections: Brazil 2014/15 to 2024/25-Long-term projections**. Available at: <http://www.agricultura.gov.br/arq_editor/PROJECOES_DO_AGRONEG OCIO_2025_WEB.pdf>. Accessed on 15 August 2016.

MARTINEZ ACOSTA, A.M.; CAYÓN SALINAS, D.G. Dinámica del crecimiento y desarrollo del banano (*Musa* AAA Simmonds cs. Gran

Enano y Valery). Revista Facultad Nacional de Agronomía Medellín, Medellín, v. 64, n. 2, p. 6055-6064, 2011.

MEDEIROS, F.A.S.B.. Relationships between growth characteristics and production of irrigated Pacovan bananas. Dissertation (Master's in Irrigation and Drainage) Federal Rural University of the Semi-Arid. 51 f. Mossoró, 2012.

MEREDITH, D.S. Banana leaf spot disease (Sigatoka) caused by Mycosphaerella musicola. Phytopathology Paper No. 11 Commonwealth Mycology Institute, 147 p. 1970.

MIRANDA, M.; ANDRADE, P.F.; CHRISTÓFORO, P.R.; COSTA, J.C. da; DIETCHFIELD, D.; GUSI, L.D.; IKEDA, P.; MACCARI JUNIOR, A.; MIRANDA, G.M.; RAMPAZZO, E.F. Study of the banana production chain in the state of Paraná. Available at: <http://www.iapar.br/arquivos/File/zip_pdf/est_cadeia_banana.pdf> Accessed on: 01 January 2018.

MOREIRA, R.S. Banana: cultivation theory and practice. 2 ed. Campinas: Fundação Cargil, 1999 335 p.

NAKASONE, J. Y.; PAULL, R. E. Tropical Fruits. 3.ed England: Biddles Ltda., 2004, p. 103-131.

NETO, A.R.; MELO, B. The banana crop. Available at: <http://www.fruticultura.iciag.ufu.br/banana3.htm#10%20%E2%80%93% 20Pr%C3%A1ticas%20culturais> Accessed on: 16 July 2018.

NÓBREGA, J.P.R.; PEREIRA, W.E.; DIAS, T.J.; RAPOSO, R.W.C.; ARAÚJO, R. da C.; OLIVEIRA, F.A. de. Pruning the pseudostem and doses of nitrogen and boron in the production of 'Pacovan' banana seedlings. Semina: Ciências Agrárias, Londrina, v. 31, s. 1, p. 1205-1218, 2010.

NOMURA, E.S.; DAMATTO JUNIOR, E.R.; FUZITANI, E.J.; OLIVEIRA E

SILVA, S. de; MORAES, W. da S. Development and production of 'Grande Naine' banana in different management systems for coexistence with Sigatoka-negra in Vale do Ribeira-SP. Revista Brasileira de Fruticultura, v. 37, p. 644-655, 2013.

NOMURA, E.S.; DAMATTO JUNIOR, E.R.; FUZITANI, E.J.; AMORIM, E.P.; OLIVEIRA E SILVA, S. de. Agronomic evaluation of banana genotypes in subtropical conditions, Vale do Ribeira, São Paulo-Brazil. Revista Brasileira de Fruticultura, Jaboticabal, v. 35, n. 1, p. 112-122, 2013.

OLIVEIRA, T.K.; LESSA, L.S.; OLIVEIRA E SILVA, S. de; OLIVEIRA, J.P. Agronomic characteristics of banana genotypes in three production cycles in Rio Branco, AC. Pesquisa agropecuária brasileira, Brasília, v.43, n.8, p.1003-1010. 2008.

PAIVA, A.; LEE, A.; KOWALSKI, M.; SILVA, O. Technical dossier: Chilling and Freesing - Injuries caused by low temperatures in tropical fruits. University of São Paulo - USP. 2011. 25 p.

PARENTE, T.V.; ARRUDA, R.J.S.; PÓVOA FILHO, N. Behaviour of 12 banana varieties (Musa spp.) in the cerrado region. Revista Brasileira de Fruticultura, Cruz das Almas, v. 3, p. 15-17, 1981.

PEDROTTI, E.L.; GUERRA, M.P.; WEIDUSCHAT, A.A. Behaviour of three banana cultivars in three planting densities. In: Congresso Brasileiro de Fruticultura, 9. 1987, Campinas. Proceedings... Campinas: Sociedade Brasileira de Fruticultura, 1988. v.1, p.147-153.

PEREIRA, M.C.T.; SALOMÃO, L.C.C.; OLIVEIRA E SILVA, de ; SEDIYAMA, C.S.; COUTO, F.A. D'Araújo; SILVA NETO, S.P. da. Growth and production of the first cycle of the 'Prata Anã' (AAB) banana tree at seven spacings. Pesquisa Agropecuária Brasileira, Brasília, v. 35, n. 7, p. 1377-1387, 2000.

PEREIRA, L.V.; OLIVEIRA E SILVA, S. de; ALVES, E.J.; REZENDE E SILVA, C.R. de. Evaluation of banana hybrid cultivars in Lavras, MG. Ciência e Agrotecnologia, Lavras, v. 27, n. 1, p. 17-25, 2003.

PEREIRA, Walter Esfrain, NÓBREGA, J.P.R.; DIAS, T.J.; RAPOSO, R.W.C.; ARAÚJO, R. da C.A.; OLIVEIRA, F.A. de. Growth and chlorophyll levels in banana seedlings as a function of pseudostem suppression, nitrogen and boron doses. Revista de Ciências Agrárias, Lisboa, v. 33, n. 2, 2010.

PEREZ, F.P.Z.; SIMÃO, S.; MANICA, I. The influence of the time of shoot selection on the development of mother plants in banana Musa Cavendishii Lamb. Cv. Nanicão. Anais...Luiz de Queiroz College of Agriculture [online].

1973, v. 30, p. 335-351, 1973.

PERRIER, X. et al. Multidiciplinary perspectives on banana (Musa spp.) domestication. Proceedings of the National Academy of Sciences of USA, Washington, v.108, n.28, p.1311-1318, 2011.

PHILLIPS, I. D. J. Apical dominance. In: WILKIDS, M. D. (Ed.). The physiology of growth and development. London: Mc Graw-Hill, 1969.

PRETI, E. A. Shoot selection time and vegetative development of the 'Grande Naine' banana tree in northern Paraná. 2014. 41p. Dissertation (Master's Degree in Agronomy) - Agricultural Sciences Centre, State University of Londrina. Londrina, 2012.

R Core Team (2016) R: A Language and Environment for Statistical Computing. R Foundation for Statistical Computing, Vienna, Austria. Available at: <https://www.r-project.org/> Accessed on: 03 Mar 2018.

RAMIREZ, Juan; JARVIS, Andy; VAN DEN BERGH, Inge; STAVER, Charles; TURNER David. Chapter 20: Changing Climates: Effects on Growing Conditions for Banana and Plantain (Musa spp.) and Possible Responses. In: Yadav, Shyam; Redden, Robert; Hattfield, Jerry; Lotze- Campen, Hermann. (Eds.) Crop Adaptation to Climate Change, Wiley- Blackwell, p. 426-438, 2011.

RAMOS, D.P.; LEONEL, S.; MISCHAN, M.M.; DAMATTO JÚNIOR, E.R. Evaluation of banana genotypes in Botucatu-SP. Revista Brasileira de Fruticultura, Jaboticabal, v.31, p.1092-1101, 2009.

RODRIGUES A.; SOUZA, A. T. de. Sobre a época de seleção dos sapiros da bananeira (Musa nana Lour.) sua desenvolvimento e frutificação na Ilha de Madeira. Agronomia Lusitana, Oeiras, v. 9, n. 2, p. 193-248, 1947.

RODRIGUES, M.G.V.; SOUTO, R.F.; OLIVEIRA E SILVA, S. de. Evaluation of banana genotypes under irrigation. Revista Brasileira de Fruticultura, Jaboticabal, v. 28, n. 3, p. 444-448, 2006.

SAMPAIO, V.R. Banana: comparative study of the Nanicão and Nanica varieties on the coast of the State of São Paulo.1967. 71 f. Thesis (Doctorate). University of São Paulo - ESALQ, Piracicaba, 1967.

SAMPAIO, V.R. Banana trees - Development and production characteristics. In: Congresso Brasileiro de Fruticultura, 4. 1977, Recife. Proceedings... Recife: Sociedade Brasileira de Fruticultura, 1978. p. 53-57.

SANTOS, S.C.; CARNEIRO, L.C.; SILVEIRA NETO, A.N. da; PANIAGO

JÚNIOR, E.; FREITAS, H.G. de; PEIXOTO, C.N. Morphological characterisation and evaluation of banana cultivars resistant to sigatoka-negra (Mycosphaerella fijiensis Morelet) in southwestern Goiás. Revista Brasileira de Fruticultura, Jaboticabal, v.28, p.449-553, 2006.

SCARPARE FILHO, J.A.; KLUGE, R.A. Production of 'Nanicão' banana under different plant densities and spacing systems. **Pesquisa Agropecuária Brasileira**, Brasília, v.36, n.1, p.105-113, jan.2001.

SCARPARE FILHO, J. A.; MINAMI, K.; KLUGE, R. A.; TESSARIOLI NETO, J. Study of the first productive cycle of the 'Nanicão' banana tree *(Musa* sp.) developed from different types of seedling. **Scientia agrícola**, Piracicaba, v. 55, n. 1. 1998.

SECRETARIAT OF AGRICULTURE AND SUPPLY, DEPARTMENT OF RURAL ECONOMY, GOVERNMENT OF PARANÁ - SEAB. **Analysis of the agricultural situation 2016/17 harvest**. Available at: <http://www.agricultura.pr.gov.br/arquivos/File/deral/Prognosticos/2017/F ruticultura_2016_17.pdf> Accessed on: 02 March 2018.

SIMÃO, S. Banana tree. In: SIMÃO, S. **Tratado de fruticultura**. Piracicaba: FEALQ, 1998. p. 327-381.

SIMMONDS, N. W.; SHEPHERD, K. The Taxonomy and origins of the cultivated bananas. **Botanical Journal of the Linnean Society**. p. 305 - 312, 1955.

SIMMONDS, N.W. **Plantains: agricultural techniques and tropical production**. 2 ed. Editorial Blume, Barcelona, Spain. 1973.

SIQUEIRA, D. L. de. **Variability and character correlations in "Prata" banana clones**. 1984. 68 f. Dissertation (Master's) - Lavras School of Agriculture, Lavras, 1984.

SONIYA, E.V.; BANERJEE, N.S.; DAS, M.R. Genetic analysis of somaclonal variation among callus-derived plants of tomato. **Current Science**, Bangalore, v.80:1213-1215, 2001.

SOTO BALLESTERO, M. **Cultivo y comercialización del banano**. 2. ed. Tibás: LIL, 1992. 649 p.

SOTO BALLESTERO, M. **Bananos: técnicas de produción, poscosecha y comercialización**. 3.ed. San José: Litografia e Imprensa LIL, 2008. 1 CD-ROM.

SUMMERVILLE, W. A. T. Studies on nutrition as qualified by development in *Musa cavendishii* Lambert. **Queensland Journal of Agricultural Science**, Queensland, v.1, p.1-127, 1944.

TAULYA, G; VAN ASTEN, P.J.A.; LEFFELAAR, P.A.; GILLER, K.E. Phenological development of East African highland banana involves trade-offs between physiological age and chronological age. European Journal of Agronomy, v. 60, n.1, p.41-53, 2014.

TEIXEIRA, L.A.J.; NATALE, W.; RUGGIERO, C.A. Changes in some soil chemical attributes resulting from irrigation and nitrogen and potassium fertilisation in banana trees after two growing cycles. Revista Brasileira de Fruticultura, Jaboticabal, v.23, p.684-689, 2001.

TURNER, D.W. Effects of climate on rate of banana leaf production. Tropical Agriculture, Trinidad, v. 48, n. 3, p. 283-287, 1971.

VALDENEBRO, J.J. de. Programmed banana harvesting. Augura, Bananeiros de Uruba, v. 10, n. 2, p. 31-40, 1984.

VAN RAIJ, B.; CANTARELLA, H.; QUAGGIO, J. A.; FURLANI, A. M. C. Fertilisation and liming recommendations for the state of São Paulo. Campinas: Instituto Agronômico/Fundação IAC, 1997. 285 p. *Technical bulletin, 100.*

VALMAYOR, R.V. Classification and characterisation of *Musa exotica, M. alinsanaya* and *M. acuminata* ssp. errans. The Philippine Agriculture Scientist, v. 84, n. 3, p. 325-331, 2001.

VICENTINI, S.; RODRIGUES, M.G.V.; SILVA, C.R.R.S. Behaviour of banana cv. Grand Naine in the south of the state of Minas Gerais. Revista Brasileira de Fruticultura, Jaboticabal, v.18, p.15-21, 1996.

VIDAL, T.C.M.. 'Nanicão Jangada' banana plant shoot selection in subtropical conditions. 2018. 65 pages. PhD thesis in Agronomy - State University of Londrina, Londrina, 2018.

WONS, I. Geografia do Paraná com fundamentos de geografia geral. 5 ed. Curitiba: Editora Ensino Renovado, 1985. 36 p.

ZONETTI, P. da C.; SANTOS, P.C. dos; BOLIANI, A.C.; SCARPARE FILHO, A.; FIGUEIRA, A.V.; SOUZA, S.A.C.D. de. TULMANN NETO, A. Evaluation of a low-growing somaclonal variant of 'Nanicão Jangada' banana *(Musa* sp.) in two densities. Revista Brasileira de Fruticultura, v. 25, n. 3, p. 471-474, 2003.

ZONETTI, P. C.; TARSITANO, M. A. A.; SANTOS, P. C.; SILVA, S. C.; PETINARI, R. A. Analysis of the cost of production and profitability of the 'Nanicão Jangada' banana tree under two cultivation densities in Ilha Solteira-SP. Revista Brasileira de Fruticultura. p. 406-410. 2002.

Printed by Books on Demand GmbH, Norderstedt / Germany